Math Guide

From Middle School to High School

By Joshua Benjamin Lee
Edited by: Dr. Perry Y.C. Lee

© 2019 Joshua Benjamin Lee and Perry Y.C. Lee
Content herein cannot be reproduced without the written consent of
Joshua Benjamin Lee and Perry Y.C. Lee

All rights reserved

Published 2019 by Amazon and Kindle Direct Publishing
Printed in the United States of America

Lee, Joshua Benjamin, 2003 –
Lee, Perry Y.C., 1966 –
Math Guide: from Middle School to High School
ISBN 978 1 6927 6969 7

Math Guide
From Middle School
to High School

$\sqrt{}$ Joshua Benjamin Lee
Edited by Dr. Perry Y.C. Lee

Acknowledgements

This project would not have been completed if it wasn't for my parents' constant reminders (nagging?) of completing what I've started. Since its inception two years ago, I am very grateful to have parents who care about what I do and the support they constantly provide to help me succeed. I would also like to thank my cousin Naomi Faith Bu for creating an amazing cover page which reflects success, not only when working with mathematics, but in every aspect of our lives.

Introduction

I wrote this book with the intention of reaching students who are like me: serious but not that serious. In part, I believe that "seriousness" comes from doing well in school, but only to the extent that school is not the single focus in your life as a middle-school or high-school student. A fair balance is needed. This is my personal journey in learning how to attain more balance, while coming to understand algebra from middle-school to high school.

I think strategically when I play video games. I also think strategically about those I compete with and those who play on my team. This has led me to think critically on how individual roles impact others and the game itself. I think of algebra as a game: a problem that needs a player to solve it, someone to win. So I think now, how can I help others who are in the position as I was in when I started algebra in middle-school? I am now in high school, and I have gone through the progression of middle-school algebra, to high school AP Calculus. So, by simply reading this book, will you be masters of algebra? Are you joking—no!! Regardless how 'smart' you are—or how smart you think you are—you still need to work consistently to maintain your grades. You have to keep in mind as well, that working smarter is more valuable than working hard. You need to be patient and invest time if you want to improve, and eventually master it. The basic knowledge of algebra will be a foundation for numerous things to come; such as helping you perform better in high school math courses, scoring well in the SAT's, and opening up more options in your pursuit of professional careers. All that hard work will pay off (I believe), and you will be very happy later that you made that investment.

Every individual learns and understands mathematics in their own way. I grew up understanding mathematics by spending time practicing similar problems that my teacher presented in the classroom during school. That's all I did. I tried to do as little as possible to get the best grades, but this is not the way to go if you want to truly understand mathematics, especially as you move towards more difficult problems. I have recently finished AP Calculus, and looking back, I realize that taking short-cuts in math has had devastating consequences in the understanding of math as I go forward. All topics in mathematics build upon previous knowledge, and often, past concepts and knowledge will pop up when you least expect it. I learned math just like the way most students did/do (by making lots of mistakes), so I will be using my personal approach and perspective on various topics in algebra, to ensure that common mistakes are avoided, and corrected in the case that they happen.

My dad is a college mathematics professor, so I asked for some of his input: "Kids now-a-days, due to technology, do not necessarily learn mathematics like the way I used to. I could be wrong, but based on my college students I had in the past and now, I see different ways to learn and understand material, for example, by watching YouTube videos. For me, when I start 'learning' a topic, it is just following a recipe, like when you first bake cookies. You can then bake the same cookies again. "Understanding" is more than "learning", which takes the learning to a better level. If you understand something, in the context of baking cookies, you can create slightly different kinds of cookies; that is, one can change the recipe without altering the taste significantly, but can bake them to taste better. In mathematics, it is just the same. Find your own way to find the correct solution. Also, I see students come to class after watching YouTube and/or Khan Academy videos, but not practicing the problems on their own. The bottom-line is that every student must work through math problems independently, through pencil and paper. Working together is also a great way to understand material as there are various ways to obtain the same solution. In addition, and equally important, asking questions and making mistakes in the classroom are important elements of understanding. Ultimately, as I mentioned earlier, every student needs to work out the solutions on their own."

Thanks for your input, Dad.

Joshua Benjamin Lee

To my parents,

*Who sparked my interest in Mathematics
so that I can use it to make sense
of the world around me*

Table of Contents

Chapter 1 Words Commonly Used and the Number System..........6
    1.1 Words Commonly Used 6
    1.2 The Number System 10
Chapter 2 The Basics 14
    2.1 The Commutative Law 14
    2.2 The Associative Law 14
    2.3 The Distributive Law 15
    2.4 Order of Operations or PEMDAS 15
    2.5 The (Dreaded) Fractions – First 'F – Word'. 19
Chapter 3 Caution!!!! 24
    3.1 Dividing by a Zero 24
    3.2 Simplifying Any Number, Especially a Fraction 24
Chapter 4 The Fundamentals – Let's Do It Correctly! 28
    4.1 'FOIL'ing-the 2nd 'F-Word'. 28
    4.2 Algebra 30
Chapter 5 The Very Important 3rd 'F –Word'. 36
    5.1 Simple Factoring 36
    5.2 Factoring Difference of Squares (or Square Numbers)...38
    5.3 Factor by Grouping 40
    5.4 Factoring (the Dreaded) Quadratic Expression 43
Chapter 6 The Quadratic Function 48
    6.1 Forms of the Quadratic Function 48
    6.2 Characteristics or Important Facts of the Quadratic Function...50
    6.3 The Quadratic Equation 54
Chapter 7 Basic Trigonometry: SOH-CAH-TOA 58
    7.1 The Sine, Cosine, and Tangent or SOH – CAH – TOA...60
    7.2 The Cosecant, the Secant, and the Co-tangent 64
    7.3 Radian Measure when using Angles 64
    7.4 The Special Right-Angled Triangles 66
    7.5 The C – A – S – T Rule 68
Chapter 8 Closing Remarks 74

Chapter 1
Words Commonly Used and the Number System

Now, let's return to the discussion that this book is about you and me. Before you and I go over various examples in algebra, I have decided there are common misconceptions or incorrect ways that my peers seem to say mathematical statements. I know because I did the same. I will iron out these misconceptions first, and I will progress through relevant topics covered in Algebra I and Algebra II that have been my school district's mathematics curriculum (which is in the state of Pennsylvania).

1.1 Words Commonly Used

Let's address 'language' or 'technical' words commonly used in mathematics.

Constant: It is simply a number. A variable 'varies' in value or can possess different values, whereas, a constant doesn't change in value. For example, 5 is a constant because 5 will remain a '5' as long as you and I are alive.

Expression[1]: It is a description of variables and constants written using multiplication, division, addition, and/or subtraction and other mathematical operations, like square roots, etc. For example, $\frac{1}{3}x^2 - 4b + 3$ and $\sqrt{x-1} - \frac{a}{a+6} + \sqrt[3]{10}$ are mathematical expressions.

---

[1] At first when I was taking algebra, I got 'expressions' and 'equations' mixed up. I thought they were the same, but as I took other math courses, I came to the realization that they are similar but different. I hope you see the difference.

Equation: It is an expression that contains an **equal** sign. So based on the two examples provided above, $\frac{1}{3}x^2 - 4b + 3 = y - a$ and $\sqrt{x-1} - \frac{a}{a+6} + \sqrt[3]{10} = 0$ are equations. Please note: An expression and an equation are different.

Factor: It is a number or an expression that is multiplied to other numbers or expressions to equal the resulting product. For example, $26 = 2 \cdot 13$, so 2 is a factor and 13 is factor of 26; $16 = 2 \cdot 8 = 4 \cdot 4$, so 2, 4, and 8 are factors of 16. Also, for the expression $x^2 + 4x - 12$, $x - 2$ and $x + 6$ are factors, since $x^2 + 4x - 12 = (x-2)(x+6)$.

Function: A mathematical relationship, when given an allowable input (called a domain), there is one output (called the range). Functions are special so they are labelled like $f(x)$ or (which reads 'f of x' or 'f at x'). For example, if $f(x) = 2x^2 - 3$ then for $x = 1$ as input, then the output is $f(1) = 2(1)^2 - 3 = 2(1) - 3 = 2 - 3 = -1$. We see that there is only one output, -1.

Polynomial: It is an expression where, in each common variable base, it has an exponent value of either 0, 1, 2, 3, or 4, etc. (any numbers 0 and above, counting numbers). For example, $x - 1$, is a polynomial: there are two terms (x and -1), and x is raised to the power of 1 (invisible), and $1 = x^0$ (x is raised to the power of a zero with $x \neq 0$). Other examples of polynomials are $b - 3b^2 + 10b^{30}$ and $y^2 - 1.2y^3 + y^{15}$, whereas, $u^{1.5} - u^3 + u^4$ is not a polynomial. Can you please tell me why?

Rational expression: An expression that is in the form of a fraction, for example, $\frac{2}{x^4 - 3}$.

Term (in an expression or equation): Parts of the expression or equation that are separated by addition and/or subtraction. For example, for the expression $\frac{1}{3}x^2 - 4b + 3$, the terms are, $\frac{1}{3}x^2$, $-4b$, and 3. Just take positive parts to the expression. In this expression, there are 3 terms.

Variable: It is commonly given a name, say 'x' or 'y'. This 'x' can possess many values, like 2, -1, 1/5, etc. Since the value of x can vary, it is called a variable.

Zero to an expression: I am **NOT** talking about the number zero, zilch, nothing, goose egg, donut hole, etc. I am talking about a mathematical word called a 'zero'. A zero to a mathematical expression means that if you plug in a number to the expression, the value of the expression is equal to zero, the actual number zero. For example, let $f(x) = x - 1$. We see that when $x = 1$, $f(1) = 0$. So, we call the number $x = 1$ the **zero** to $f(x)$.

Practice Problem Set #1

1. Given $s - \dfrac{x}{y-1} + (x-1)z^2 + 3$, how many terms? What are the terms? Also, is this an equation or an expression? Identify the term(s) that are variables and constants.

2. Find the zero(s), if any, to $y = g(x) = x^2 - 4$.

3. Please provide two examples of a variable and two of a constant that you see or deal with in your daily life.

1.2 The Number System

Different Classes/Types of 'Numbers'

In mathematics, there are different classes of numbers and they are mentioned and assigned proper labels to these types of numbers.

i) <u>Natural Numbers</u> (or counting numbers): 1, 2, 3, ..., 100, 101, 102, 103, 104, ... and so on.

ii) <u>Whole Numbers</u>: A set of Natural numbers with a zero (0). That is, 0, 1, 2, 3, 4, ... and so on.

iii) <u>Integers</u>: a set of natural numbers, the zero, and the set of 'negative' natural numbers. For example, ... -1235, -1234, ..., -2, -1, 0, 1, 2, ..., 556, 557, 558, ..., 1000023, 1000024, ... and so on.

Rational Numbers and Irrational Numbers

iv) <u>Rational</u> means 'fraction' or 'integer over integer. For example, 1/3, -4/9, 5, -201. Five (5) is also rational number because 5 = 5/1. All integers are rational numbers.

v) <u>Irrational</u> means numbers that **CANNOT** be written as a rational number or a fraction. For example, $\sqrt{2}, \pi$, and $\sqrt[3]{15}$ are rational numbers. Please note that irrational numbers, when written in decimal form, go on forever **without** a definite pattern.

vi) <u>Real Numbers</u>: They are comprised of a set of numbers that are rational and irrational numbers (mentioned above).

vii) <u>Complex Numbers</u>: A set of numbers that are made up of two numbers (parts): the real number and the imaginary number. For example, *standard* form of a complex number is given by -2 + 3*i*, where $i = \sqrt{-1}$. The real (number) part of the complex number is -2, and the imaginary (number) part of this complex number is $3i$. Another example of a complex number is $2.345 - i\sqrt{3}$. In this case, the real part is 2.345 and the imaginary part is $-i\sqrt{3}$.

Practice Problem Set #2

1. Classify each number given:

-10.1, $\sqrt[3]{10}$, $\dfrac{\pi}{3}$, $\dfrac{12}{2.5}$, $\dfrac{3}{\sqrt{\pi}}$, $\dfrac{1}{1-i}$, $\dfrac{\sqrt{100}}{2}$ and $-\dfrac{500}{3}$.

2. Please provide examples of real numbers that a banker would deal with? How about an auto mechanic? Doctor, say, a surgeon? Police officer?

3. Is a real number a complex number? For example, -2.5, is this a complex number?

Chapter 2
The Basics

Laws in Algebra

I find laws in algebra to be obvious, but despite being obvious, I abuse it quite often. Let's look at the three laws that we all should (intuitively) know. Also, let's review the order of operations, also known as PEMDAS.

2.1 The Commutative Law

It states that given any two numbers, say x and y, **under addition**, $x + y = y + x$. That is, for example, $2 + 3 = 3 + 2$. Yeah, obvious! Also, under multiplication, $x \cdot y = y \cdot x$. That is, $2 \cdot 3 = 3 \cdot 2$, which is also obvious!! But **under subtraction**, the **commutative law does not apply**! That is, $2 - 3 \neq 3 - 2$, similarly, under division $2/3 \neq 3/2$.

2.2 The Associative Law

This law states, given three numbers x, y, and z, $(x + y) + z = x + (y + z)$. Another way to state this law is that if there is a series of additions, how you group them under addition does not matter, as long as you add each group correctly. That is, $(2 + 3) + 4 = 2 + (3 + 4)$, which says $5 + 4 = 2 + 7$ is correct! This law also applies under multiplication: $x \cdot (y \cdot z) = (x \cdot y) \cdot z$. So, $2 \cdot (3 \cdot 4) = (2 \cdot 3) \cdot 4$ which is $2 \cdot (12) = (6) \cdot 4$ is correct!

Pretty 'obvious', isn't it? It is, once you practice and practice!!

2.3 The Distributive Law

The distributive 'law' in algebra is one law that is used often in algebra. Here is an example illustrating the distributive law: $2\cdot(3+7) = 2\cdot 3 + 2\cdot 7$. Number 2 is distributed inside the parenthesis (which is a sum) under multiplication; that is, 2 is distributed to 3, and 2 is distributed to 7, and then added. So using variables, the distributive law is written this way: $a\cdot(b+c) = a\cdot b + a\cdot c$.

2.4 Order of Operations or PEMDAS

There are four basic operations in mathematics (addition, subtraction, multiplication, and division). The process of calculating a series of these operations (that is, doing arithmetic) is to follow a set of 'rules' or guidelines by an acronym that we all should know by PEMDAS – commonly known as 'Please Excuse My Dear Aunt Sally'. It stands for the order to follow when doing a series of arithmetic operations: do **P**arentheses (what's inside) first – then **E**xponents – then **M**ultiplication or **D**ivision (as they appear *from left to right*) – and then **A**ddition or **S**ubtraction as they appear *from left to right*.

Example #1: Evaluate: $-2 + 3\cdot 6 - 14/2\cdot(2 - 3\cdot 2^3)$

Solution:
I followed order of operations, and based on my scribbles, my answer is 160. Please let me explain.

Clarification to the solution:
Do what's inside the parenthesis first (P). Then again, follow PEMDAS inside this parenthesis: $2 - 3\cdot 2^3$. Do exponents first: $2^3 = 2\cdot 2\cdot 2 = 8$. So, $2 - 3\cdot 2^3 = 2 - 3\cdot 8$. Now multiply $3\cdot 8$, that is, $2 - 3\cdot 8 = 2 - 24$. Then, the subtraction $2 - 24 = -22$. Now back to the overall problem: $-2 + 3\cdot 6 - 14/2\cdot(2 - 3\cdot 2^3) = -2 + 3\cdot 6 - 14/2\cdot(-22)$. Note that the value inside the parenthesis is -22 (as shown). We now follow the order of operation for the overall problem; that is, do the multiplication and

division as they appear *from left to right*: $-2 + 3 \cdot 6 - 14/2 \cdot (-22) = -2 + 18 - 7 \cdot (-22) = -2 + 18 - (-154) = -2 + 8 + 154 = 6 + 154 = 160$. As you can see above, I delineated each operation by following PEMDAS. Note: for the operations involving $-14/2 \cdot (-22)$, I followed the rule 'multiplication or division as they appear from left to right'. So, since $-14/2$ came first, it is -7 then multiply it with -22. It would be **incorrect** to do $2 \cdot (-22)$ first then divide that result into -14. The same rule applies when you have a series of additions and subtractions: perform them as they appear from 'left to right'. Not following PEMDAS will allow your calculations to be incorrect and can be disastrous (for the world around us).

Example #2: Evaluate: -1^3.

Solution:
The answer is -1. I remember my dad asking me this question. He said to explain how I got the *correct* answer. I said this: $-1^3 = (-1)(-1)(-1) = -1$. The way he looked at me, I think he meant I got away with a serious crime! Coincidently, I got it correct, but *how* I got the correct answer is *incorrect*!! This I remember, he said, "it is not destination but the journey". What the ...? I just got lucky that I got it correct. In math, there is some luck involved when getting the answer correct, but that's not what understanding mathematics is about. In fact, there is a lot of frustration, but once I got to understand certain topics, it was well worth my time. Let me explain why I did it incorrectly.

Solution Version 2.0:
Please note that -5 is the same as $(-1)(5)$. So, using this approach, $-1^3 = (-1)(1^3) = (-1)(1) = -1$. Please note: -1^3 is not the same as $(-1)^3$, which I assumed (which is incorrect!). If it is, it would have been written that way!! Or another way to look at it is, according to PEMDAS, exponent is done first here (since there is no parenthesis) which 1 is raised to the power of 3. Then this value is negated; that is, multiplied by -1. That is, $-1^3 = (-1) \cdot (1)^3 = (-1) \cdot (1 \cdot 1 \cdot 1) = -1$.

Example #3: So my dad continued with a similar problem and asked me to do this: -2^4.

My Solution:
Based on the previous example, the answer is -16. We know from the previous example, $-2^4 \neq (-2)^4$!! That is, $-2^4 = (-1)(2^4) = (-1)(2 \cdot 2 \cdot 2 \cdot 2) = (-1)(16) = -16$. Do exponents first before multiplying by -1 to it.

Aside: He over-reacts!! Who cares? I think he takes it personally, and I think I know why: math is his 'bread and butter' for him or 'kimchi and bap' (for those who are Korean). After several years, I discussed this incident with my dad, and asked him why he reacted to me the way he did. His response was something like, 'couldn't believe that I was the same genetic makeup as him' or something similar to those words. Well, he does not look like me ... I look more like my mom.

Example #4: $10/2 \cdot (-1 + 4)$. The answer is 15. Using PEMDAS, we calculate what's inside the parenthesis first: $-1 + 4 = 3$. So $10/2 \cdot (-1 + 4) = 10/2 \cdot 3 = 5 \cdot 3 = 15$. Although the acronym is "MD", it is multiplication or division **whichever comes first** as you do the operations from left to right. So do the 10/2 first (division) then afterwards, multiply that result (which is 5) with 3, which is 15.

Practice Problem Set #3

1. Simplify: $3^3 / 2 \cdot 5 - 4 / 3 \cdot 3^4 / (10 - 2 \cdot 3) + 1$

2. Would you use PEMDAS or the distributive law to simplify $-3(\frac{4}{3} - 3 \cdot 2)$. Please explain why you have chosen what you chosen.

2.5 The Dreaded Fractions - First 'F -Word'

The word 'fraction', is one of three 'f – words', which will appear throughout this book. We cannot avoid fractions – like I cannot avoid my parents always telling me this and that. We will review the four operations involving this 'f – word'.

Before we start, I am going to be blunt: I hate fractions. My dad says he doesn't like fractions either. He says it just comes with the math territory.

My dad said this to me on several occasions: "10 out of 3 students don't like fractions". I didn't find it funny on those occasions nor do I find my dad funny.

2.5.1 Multiplying Fractions

Multiplying fractions is the easiest: simply multiply across ... numerator times numerator and denominator times denominator. Let's look at this problem: simplify $\frac{3}{8} \cdot \frac{16}{9}$.

When multiplying or dividing fractions, NO common denominator is required!! I used to find a common denominator and then multiply across, but this is completely **NOT** necessary and a complete waste of time and effort! There are two ways to approach this. One approach is to simply multiply 'straight' across. After, the fraction must be simplified!! You remember? Always simplify the answer!

So, $\frac{3}{8} \cdot \frac{16}{9} = \frac{48}{72} = \frac{6 \cdot 8}{6 \cdot 12} = \frac{\cancel{6} \cdot 8}{\cancel{6} \cdot 12} = \frac{2 \cdot \cancel{4}}{3 \cdot \cancel{4}} = \frac{2}{3}$. The other and a much better (efficient) approach is to factor out the numbers and cancel some of them first before multiplying across.

That is, $\frac{3}{8} \cdot \frac{16}{9} = \frac{3}{8} \cdot \frac{2 \cdot 8}{3 \cdot 3} = \frac{\cancel{3}}{\cancel{8}} \cdot \frac{2 \cdot \cancel{8}}{\cancel{3} \cdot 3} = \frac{2}{3}$ which is the same answer from above. In this latter case, there was no multiplication required.

Note: It does not matter how you do it, as long as it is done properly. My dad says this to me a lot: "It is not the destination that is important, but the journey". Whatever!

2.5.2 Dividing Fractions

Dividing two fractions involves a similar approach for multiplying two fractions with an extra 'manipulation': invert the bottom fraction (divisor), and multiply it with the fraction that is on the numerator. For example, $\frac{2}{3} \div \frac{1}{5} = \frac{2/3}{1/5} = \frac{2}{3} \cdot \frac{5}{1} = \frac{2 \cdot 5}{3 \cdot 1} = \frac{10}{3}$. Leave the fraction as it is (as an improper fraction), and if possible, simplify. That is, do not write it in 'mixed' form: $3\frac{1}{3}$. My dad tells me my grandfather wrote fractions this way, and I shouldn't as times have changed.

2.5.3 Adding and Subtracting Fractions

Adding and subtracting require some attention. So be careful. It involves finding a 'common denominator' based on the denominators of two fractions. Let's look at an example: $\frac{2}{3} - \frac{1}{5}$. A common denominator is a number that is both divisible by both 3 and 5. We see that 15 (= 3·5) is a common denominator. For the first fraction ($\frac{2}{3}$), this fraction is rewritten so that the denominator is 15. Since initially, the denominator of this fraction is 3, to get to 15, the denominator needs to be multiplied by 5. To maintain the same value of 2/3, the numerator is also multiplied by 5. So $\frac{2}{3} = \frac{2 \cdot 5}{3 \cdot 5} = \frac{10}{15}$ and similarly for the other fraction $\frac{1}{5} = \frac{3}{15}$. Since we now have a common denominator, so simply subtract the numerator and keep the denominator at 15. That is, $\frac{10-3}{15}$. Here is the summary of what I just said: $\frac{2}{3} - \frac{1}{5} = \frac{2 \cdot 5}{3 \cdot 5} - \frac{1 \cdot 3}{3 \cdot 5} = \frac{10}{15} - \frac{3}{15} = \frac{10-3}{15} = \frac{7}{15}$. In this case, the answer cannot be simplified any further.

Practice Problem Set #4

1. Simplify: $\frac{3/4}{2/9} \cdot \frac{10}{3} \div \frac{7}{6} - \frac{1}{2}$

2. Simplify: $(3 - \frac{1}{4})(-1 + \frac{10}{5})$

3. Simplify: $(3 - \frac{1}{4}) \div (-1 + \frac{10}{5})$

Chapter 3

Caution!!!

Based on my experience thus far in taking high school mathematics (Calculus AP), I have found that there are two 'no-no's in mathematics.

3.1 Dividing by a Zero

Dividing by a zero is a 'no-no' in mathematics! For example, two divided by zero is NOT zero!! It is called 'undefined' (we don't know what it is; it is also called infinity). Try that on your calculator and see what is displayed … 'error'!! My dad says that dividing by a zero is the 'ultimate sin' in math.

3.2 Simplifying Any Number, Especially a Fraction

My English teacher told me when writing a grammatically correct sentence, make sure it is clear and simple enough so that any reader understands it. Isn't that what communication is about? As a human being, it is customary and important that when we convey a message to someone (no matter how technical it is), we need to simplify it so that the intended parties understand. In a similar note, when we convey a number to someone, that number needs to be simplified. For instance, my parents tell me that the Lee family has three children. He did not say '$9/3$' children. Also, when someone asks you how old you are, you need to say a number, not a convoluted number. I just turned sixteen, but I would not dare to say '$2 \cdot 7 + 2$' nor '$320/20$'. Although both are exactly sixteen, we do not communicate our age (or any other number) this way. If carpenters communicated numbers this convoluted way, any simple job will not be completed or the work will be completely messed up, and a lot of time will be spent saying 'what?' for clarifications. In mathematics, we simplify our numbers when we present them, and not doing so will also make the person look at times foolish. So, simplify your answer!!!

Practice Problem Set #5

1. Find $\dfrac{2}{0.1}, \dfrac{2}{0.01}, \dfrac{2}{0.001}, \dfrac{2}{0.0001}, \dfrac{2}{0.00001}$, and $\dfrac{2}{0.000001}$.

Based on your result, please explain why when you divide by zero, the answer is 'undefined' or a very large number.

2. Under what situation would you NOT simplify your answer. Why?

Chapter 4

The Fundamentals – Let's Do It Correctly!

I am sure you have heard of the acronym FOIL. This acronym is essentially the distributive law, but educators came up with this acronym so that students can easily remember it.

4.1 'FOIL'ing– the 2nd 'F – Word'

'FOIL' is the 2nd 'f - word' out of three 'f – words' presented in this book.

Let's focus now on expanding/'FOIL'ing $(a+b)^2$. I used to write this: $(a+b)^2 = a^2 + b^2$ because I was in a hurry (for what?). Is this correct? **FOIL** is an acronym for students to memorize when multiplying two, two-termed expressions. Since $(a+b)^2 = (a+b)(a+b)$ let's use FOIL to expand this.

If you forgot, '**F**' stands for 'first', '**O**' stands for 'outer', '**I**' stands for 'inner', and '**L**' stands for 'last'. So take one term from each of the two expression (in the parentheses) and multiply. That is, multiply the first terms: 'a with a'; Then multiply the outer terms: 'a times b', inner terms which is 'b times a', and 'last', 'b times b'. In a nutshell, all you are doing is multiplying all but different terms together.

So, $(a+b)^2 = (a+b)(a+b) = a^2 + ab + ba + b^2 = a^2 + 2ab + b^2$.
As you can see, $(a+b)^2 \neq a^2 + b^2$!! Also, plug $a = 1$ and $b = 5$ to see if this is true: $(a+b)^2 = a^2 + b^2$. If this is true, then you discovered something new in mathematics, and you must bring this forth to the

SLMLN (which you should know, it's the **Supreme Leaders** of the **Math League of Nations**).

Practice Problem Set #6

1. Multiply: $(x-y+1)(x+y-1)$, $(2a+y)(a-y+1)$

2. Expand: $\dfrac{1}{a}(1-a-a^2-a^3-a^4)$

4.2 Algebra

There is a saying, 'curiosity killed the cat'. Likewise, there's probably a quote like, 'not doing algebra properly ruined the life of a science or math student'. This is true (based on my limited existence on earth) since knowledge of algebra is a requirement in the sciences and obviously in mathematics.

Let's look at common mistakes that I had made in the past, and at times, I still continue to make when I get careless.

4.2.1 Cancelling a Variable (Improperly)

We start off with a typical example that some students make. Let's simplify a rational expression like this: $\frac{x-1}{x-2}$. In fact, you cannot simplify this any further. Let me explain.

A mistake that I made earlier (in my career as a student) is cancelling the variable x in the numerator and the denominator like this: $\frac{x-1}{x-2} = \frac{\cancel{x}-1}{\cancel{x}-2} = \frac{-1}{-2} = \frac{1}{2}$. Please note that based on this cancellation technique, this rational expression simplifies to a constant equal to ½. Rather, if we substitute $x = 3$, then it simplifies to $\frac{3-1}{2-1} = \frac{2}{1} = 2$. So, does $\frac{1}{2} = 2$? Well, no. So, just be careful. The cancellation that was made defies the laws of mathematics, and you just cannot do that because the numerator contains 'subtraction' and similarly in the denominator.

Advice: <u>Cancelling a common variable or constant can only occur when that variable or constant appear as multiplication in both the numerator and denominator.</u> Let's clarify these concepts using another example: $\frac{5a}{10a}$. We know that '$5a$' is the same as '5 times a' and similarly, '$10a$' is '10 times a'. So, the "a's" (a common variable which appears in both the numerator and the denominator) cancels out and simplifies to $\dfrac{5a}{10a} = \dfrac{5\not{a}}{10\not{a}} = \dfrac{5}{10} = \dfrac{1}{2}$ (as long as a is not equal to 0 – remember that you cannot divide by a zero!). Please note that after the cancellation, the fraction is simplified to 1/2. So, if you plug in any value of a (other than zero), this value will always equal ½.

4.2.2 Using the Distributive Law Improperly

Let's look at an example that I did.

Example #5: Let's do a common problem that we've all seen before. Solve a linear equation (for x): ½ (x -1) = x + 2. You can guess for the value of x, but this approach will be slow, and possibly, you will not be able to find the solution efficiently. So to do it 'efficiently', we use the laws of mathematics addressed in Chapter 2 to find x.

Here is my solution:

There is a fraction in the equation. To get rid of the fraction (that is, the '2' in the denominator), multiply both sides (on the left side and on the right side) by 2. You remember that whatever operation you perform on one side of an equation, you must do the same operation on the opposite side of the same equation. So, $2\left[\frac{1}{2}(x-1)\right] = 2(x+2)$ which simplifies to $x - 1 = 2x + 2$ or $-3 = x$. So the solution is $x = -3$. My dad asked, "how do you know if this $x = -3$ is correct?" You need to check whether you or I did it or anybody did it correctly. So off I checked: left side (LS) = $\frac{1}{2}(-3-1) = -2$ and the right side (RS) = $-3 +2 = -1$. Note that LS $\neq$ RS, so no, $x = -3$ is NOT the solution!! That's why it is wise to always check!! Where did I mess up? Please revisit what I did wrong. Do you see where I made the mistake? If you look at the RS of the equation above, $2(x + 2) = 2x + 4$, NOT $2x + 2$!!!

I applied the distributive law incorrectly!! Didn't I say the laws look obvious? I guess not. So, $x - 1 = 2x + 4$ which is $x - 5 = 2x$ or $-x - 5 = 0$ or $-x = 5$ or finally $x = -5$. You can check whether this is the correct answer!

4.2.3 Breaking the 'law' when using exponents. Please don't!!

Here are the Laws of Exponents:

i) $x^a x^b = x^{a+b}$; for example, $2^5 \cdot 2^7 = 2^{5+7} = 2^{12}$

ii) $\dfrac{x^a}{x^b} = x^{a-b}$; for example, $\dfrac{15^7}{15^3} = 15^{7-3} = 15^4$

iii) $(xy)^a = x^a y^a$; for example, $(3 \cdot 7)^4 = 3^4 \cdot 7^4$

iv) $\left(x^a\right)^b = x^{a \cdot b}$; for example, $\left(23^5\right)^6 = 23^{5 \cdot 6} = 23^{30}$

v) $x^{-a} = \dfrac{1}{x^a}$; for example, $21^{-4} = \dfrac{1}{21^4}$

Example #6: Simplify $\left(2x^4\right)^3$.

I will quickly provide the answer without showing you the steps. The answer is $\left(2x^4\right)^3 = 2x^{12}$ Is it correct? Quick recap: exponents are used to conveniently articulate constants and/or variables that are multiplied repeatedly. Having said this, $\left(2x^4\right)^3$ is rewritten as repeated multiplication: $\left(2x^4\right)^3 = \left(2x^4\right)\left(2x^4\right)\left(2x^4\right) = 2 \cdot 2 \cdot 2 \cdot x^4 \cdot x^4 \cdot x^4 = 8x^{4+4+4} = 8x^{12}$. Notice how the constants and variables are 'regrouped' based on the associative law under multiplication. As a result, the '2' is multiplied 3 times just like x^4. So, don't forget to raise 2 to the power 3 as well (from part iii in the Laws of Exponents above), which I didn't do when I did it quickly. So don't break any laws ... or else!!

Practice Problem Set #6

1. Simplify: $\left[\left(\dfrac{2x^{-3}y^4}{3x^2}\right)^{-2}\right]^{-5}$ and $\left(\dfrac{3ab^4c^2}{a^{10}b^{-2}c^5}\right)^{-2}$

When simplifying, please make sure that the exponents are all positive.

2. Solve for a: $\dfrac{2}{3}(1-a) = \dfrac{4a}{3}\left(\dfrac{1}{a} - 1\right)$

Chapter 5

The Very Important 3rd 'F – Word'

Factoring is the last 'f – word' from a total of three 'f – words'.

There is no 'formula' when it comes to factoring. There are different ways to factor but there is no 'cookie-cutter template' or a direct once-for-all formula as algebraic expressions come in many different forms. Let's look at various strategies/methods.

5.1 Simple Factoring

What I mean by 'simple' factoring is to factor out the greatest common factor (GCF) from the expression. To do this, look at each term of the expression and see what the GCF is. What is a GCF? It is the 'greatest' or largest number/variable that can be divided into by each term in the expression. Okay, say you have two numbers 12 and 32, the GCF is 2? No since 4 can also divide into 12 and 32. How about 5, 6, ... etc.? We see that the GCF is 4. There is no other number higher than 4 that can divide into 12 and 32 evenly.

Example #7: Factor $2x^2y - 4xy^2 - 10xy$

Look at each term and find the GCF. We see that 2 is a common factor, along with x and y, so the GCF is $2xy$. When you factor out the GCF, $2xy$, you get: $2x^2y - 4xy^2 - 10xy = 2xy(x - 2y - 5)$. Simple. How do you check that you did the factoring correctly? Distribute $2xy$ back into the terms in the parentheses ... and that is known as the distributive law! Always check your work!! Checking your work should and must be a habit, and it should always be a part of your life.

Problem Exercise Set #8:

Factor: $24a^3b^7xy^4 - 3abxy^2 - 12abcxy$ and $4x^2 - 12y^2 - 10xy$

5.2 Factoring Difference of Squares (or Square Numbers)

What is a difference of squares? It is literally the difference of square numbers!! For example, 4 – 9 is a difference of square numbers. Why? 4 is written as 2^2 and $9 = 3^2$. So, $4 - 9 = 2^2 - 3^2$ which is a difference of squares. The difference of squares has very unique factors, in fact two similar factors.

Let's consider an example: factor $a^2 - b^2$. As you may know, it is $a^2 - b^2 = (a+b)(a-b)$. Factors of difference of squares appear as 'conjugate' pairs. That is, one factor is $a + b$ and the other is the difference $a - b$. That's it!! Let's do another example since it's so fun!

Factor $4x^2 - 1$. How is $4x^2 - 1$ a difference of squares? Need to rewrite $4x^2$ and 1 as squares!! We know that $4x^2 = (2x)^2$ and $1 = 1^2$. So $4x^2 - 1 = (2x)^2 - (1)^2 = (2x+1)(2x-1)$. I realize that in math, at times, you must fit things into a 'cookie cutter' template. This example is just that. It is important that at times, you may need to recognize an expression as a difference of squares when asked to factor an expression.

Practice Problem Set #9

Factor completely: $3x^2 - 6$, $9x^2y^2 - 49$, and $y^2 - 5$. For the last problem, here's a hint: you can use square roots.

5.3 Factor by Grouping

Some of you may not have heard of this method/technique. Well, it has been around much longer than you and me, and I guess, it does not get that much publicity. I did not know about this method initially, but was introduced to it later. It is a *robust* method yet not many students know about it. Learn it and you'll be happy that you know it. It's like a diamond in the rough, so they say. So let's factor a polynomial using this technique. If you forgot what a polynomial is, you will know just by looking at it or see Chapter 1, Section 1.1. It's just a fancy word for a particular form of an expression.

Example #8: Factor: $4 - 4x^2 - x + x^3$. This is a polynomial.

Whenever you see four terms in a polynomial expression, try factoring it by grouping first. Also, rewrite/regroup each term in the polynomial in decreasing powers of x if it is not. That is, we need to regroup the four-termed polynomial to this: $x^3 - 4x^2 - x + 4$

Now, the first step is to separate the four terms into two groups of two terms: first group is "$x^3 - 4x^2$" and the second group is "$-x + 4$".
For each group, factor out the greatest common factor (GCF). So, $x^3 - 4x^2 - x + 4 = x^2(x-4) - (x-4)$. We notice here that for the second group, "-1" was factored out; that is, $-x + 4 = -(x-4)$. Always do that when you see a 'negative' of the variable in the second group. The reason is, this way, the factor $x - 4$ appears in both the first group and the second group.

So what just happened when we split the original polynomial into two groups and factored each group? The four terms in the polynomial changes to two terms. They look different but they are the same polynomial

Advice: Smaller the number of terms in an expression, easier it is to factor. Since this factor is common to the two terms, you can factor that out: $x^2(x-4)-(x-4) = (x-4)(x^2-1)$. So, are we done factoring? We're not done yet because (x^2-1) can be factored further. This is in the form of a difference of squares (from before), so $(x^2-1) = (x+1)(x-1)$. So, $x^3 - 4x^2 - x + 4 = (x-4)(x+1)(x-1)$. Here we factored the polynomial *completely*.

Observation: It is fascinating that in math, you may need previous material to help and use the current method. I am still fascinated how math builds from previous material/knowledge! The more you know, the more power to you!!

Practice Problem Set #10

Factor completely (which means factor it as much as you can): $2x^3 - 8x^2 - x + 4$ and $y^3 - 2y^2 + 4y - 8$.

5.4 Factoring (the Dreaded) Quadratic Expression

A quadratic expression in general look like this: $ax^2 + bx + c$, where a, b, and c are real numbers.

A quadratic expression is also called a polynomial of *degree* two. The degree refers to the highest power that appears in the polynomial. For example, $-2x^2 - 3x + 10$ and $10y^2 + 4y - 1$ are both quadratic expressions or polynomials of degree 2. If you want to be hip, just say quadratic expression!

Please note that a cannot be zero because if it is, the expression is a linear expression. Factoring linear expressions is much, much simpler than factoring quadratic expressions.

Why is there so much emphasis on factoring in high school? I think factoring is like summarizing a mathematical expression, so that later, easier computations can be achieved. It's like writing a news article as opposed to writing a story in a novel. Factoring will allow a mathematical expression to be articulated in a concise manner, and also can see the important parts of the expression like the factors (if it can be factored). Let's factor a quadratic expression.

Example #9: Factor $x^2 - 6x - 7$.

Factoring requires *informed* guessing. You must use information provided to you from the quadratic expression, and make 'guesses' until these 'guesses' work out.

Try the first three factoring methods discussed above. They don't work on this example. Please verify my statement by trying the first three methods yourself.

To factor the quadratic expression, make sure you identify the value of a, which is $a = 1$. Then, identify the number multiplied to x (called the coefficient of x) which is $b = -6$. Also, identify the constant term, $c = -7$. To factor this expression a ***product-sum*** idea is used. We use these two numbers, -6 and -7, which are the clues that you will use to factor the quadratic expression. So based on these two numbers, ask "what two numbers when I ***add***, it is -6, and at the same time when I ***multiply*** them, it is -7?" The best way do this is by trial and error (yes!!), by first finding the integer factors of -7. List them: they are 1 and -7, and -1 and 7, any others? No. So which pair when you add them will equal -6? The pair is 1 and -7. So the factors are $(x + 1)$ and $(x - 7)$, which means $x^2 - 6x - 7 = (x+1)(x-7)$. FOIL it to check!! It is wise to check, and check, and check.

Note: Factoring a quadratic expression is the reverse process of FOILing - I call it 'unFOIL'ing. Un'FOIL' to check whether you factored it correctly.

Let's do another example, similar to this one but with a twist.

Example #10: Factor $2x^2 + 5x - 7$.

Do you see the difference between this example and the previous one (other than being a different quadratic expression)? If we use the same approach as above, let's see whether this 'product-sum' idea works here. As in the previous example, we take the coefficient of x (which is $b = 5$), and the constant term (which is $c = -7$). So once again, let's see if there are there two numbers so that the sum is 5 and the product -7? List the factors of -7: -7 and 1, and 7 and -1, that's all. We see that the sum of those factors does not add up to 5. So, does that mean $2x^2 + 5x - 7$ cannot be factored? Well, not really. We need to look closely that in this quadratic expression, the number attached to x^2 (which is the coefficient) is $a = 2$. In the previous example, the coefficient of x^2 is 1 so the product-sum approach works smoothly. Since the coefficient is not equal to 1, a slight adjustment needs to be made, but the same 'product-sum' approach is also used. In this case, multiply the coefficient of x^2, which is $a = 2$, with the constant term, $c = -7$: it gives $(2)(-7) = -14$. So take this product (-14) and the coefficient of x (which is $b = 5$), and ask the same question like before: what two numbers when you multiply them, you get -14 and at the same time when you add them, 5? List the factors of -14: -14 and 1, 14 and -1, 2 and -7, and -2 and 7. It is clearly -2 and 7. Great! But don't be in a rush to say $(2x-2)(x+7)$ is the factored form of $2x^2 + 5x - 7$??!! FOIL it to see... and you see that it isn't so! What did I say? Don't be in a hurry and check your work!

What we now do is take those two numbers -2 and 7, and write the original quadratic expression this way: $2x^2 + 5x - 7 = 2x^2 - 2x + 7x - 7$. Do you see what I did here? I rewrote $5x$ as $-2x + 7x$. That is, take the middle term ($5x$), and use the two numbers identified above to rewrite the middle term as a sum and multiplied by x for each. That is, $5 = -2 + 7$ or $5x = -2x + 7x$. That's it! We now finish the factoring process by using the *method of factor by grouping*. So,
$2x^2 + 5x - 7 = 2x^2 - 2x + 7x - 7 = 2x(x-1) + 7(x-1) = (2x+7)(x-1)$
Always check by FOILing.

Observation: We actually used previous knowledge (factor by grouping) to complete the factoring process when $a \neq 1$.

Summary: So if $a \neq 1$, split the middle-term (with variable x) into two terms as a sum with the identified two numbers, and proceed by factor by grouping.

Practice Problem Set #11

Factor: $4x^2 - x - 3$, $2a^2 - a - 6$, and $-x^2 + 3x + 28$.

Chapter 6
The Quadratic Function

Quadratic expressions and quadratic functions are subtly different but similar. A quadratic function is based on a quadratic expression that is calculated and assigned to an output, say a y-value which has an input say, x-value. As a result, quadratic functions can be plotted (or graphed) on the Cartesian Coordinates since a set of ordered pairs, (x, y) can be generated/calculated.

I believe you have heard of the quadratic function, which is also called the *parabola*. I will use the quadratic function and the parabola interchangeably. When I was first introduced to this quadratic function and did examples related to it, I did not realize the importance of this function, and where this function appears (for example, in physics).

Let's look at the important features of quadratic functions.

In this chapter, we will:

- Present two different forms of the same quadratic function;
- Identify the shape of the quadratic function;
- Locate the vertex of a quadratic function; and
- Determine where the quadratic function crosses the y-axis, and if at all, the x-axis.

6.1 Forms of the Quadratic Function

There are two ways (or forms) to mathematically represent a quadratic function. It is represented either by:

- $f(x) = ax^2 + bx + c$, called a **_standard_** form; or by
- $f(x) = a(x-h)^2 + k$, called a **_vertex_** form, the vertex being (h, k)

They are the same quadratic function, but articulated differently. The values of a, b, and c, and h and k, noted above provide important geometric information, namely, the shape of the parabola.

The vertex form is often used when you want to 'sketch' the quadratic function without finding a bunch of points (ordered pairs). What I mean by a 'sketch', I mean to provide a good but an approximate representation of the exact shape of the parabola. That is, it does not have to be perfect. Also, I will show you how you can switch from the standard form to the vertex form, and vice versa. This involves, as always, good knowledge of arithmetic and algebra. Always check your work afterwards!!

A sketch of the parabola on the Cartesian co-ordinates will show that the parabola either 'opens up' or 'opens down'. Why can't a quadratic function open sideways (to the right or to the left)? It's because ... please explain clearly!

Fact #1: There is an important value 'a' which is called the '*coefficient*' of x^2. It is simply the number 'attached' to x^2. This value of 'a' determines whether the parabola 'opens up' or 'opens down', which determines the fundamental shape of the parabola. If the values of a are positive ($a > 0$), then the parabola 'opens up'. If a is negative ($a < 0$), then the parabola 'opens down'. It is just a matter of identifying (looking at) the value of a once the quadratic function is provided.

Also, there is a point on the parabola that is called a 'vertex', which is a very important feature of the parabola which has several interpretations. What is a vertex? You've heard of the word 'vertex' in geometry. It is where two sides, for example, of a triangle meet at a point or also called a corner point. Also, a vertex can be found in a room, where the three walls meet at a corner (also called a vertex). Similarly, the two 'sides' (curved sides) of a parabola meet at a point also called the vertex.

Fact #2: A vertex of a quadratic function, when plotted on the Cartesian coordinates, is either the **_lowest point or the highest point_** (h, k) on the parabola, depending on whether the parabola either 'opens up' or 'opens down', respectively. This vertex (h, k) can also be interpreted as at $x = h$, the **_maximum or the minimum_** value of the quadratic function is $y = k$.

Let's look at an example of a quadratic function to demonstrate what I just talked about.

Example #11: Consider $y = f(x) = x^2 - 2x - 24$ and let's sketch it.

Here, the quadratic function is written in standard form. First we identify the simple stuff: $a = 1$, $b = -2$, and $c = -24$. Each of these values provides some 'clues' to the shape and the position of the quadratic function in the Cartesian co-ordinates. Let's consider the importance and the significance of the value 'c', and in this case, it is $c = -24$, the constant term.

Fact #3: What happens when you substitute $x = 0$ to $f(x)$, that is, evaluate $f(0)$? We can easily see that $y = f(0) = 0^2 - 2(0) = -24$, which is the location where the parabola crosses the y-axis. So the location where the parabola crosses the y-axis, as an ordered pair, is (0, -24), also called the **_y-intercept_**!! So, in general, (0, c) is the location, as an ordered pair, where the quadratic function crosses the y-axis.

Now let's rewrite this quadratic function in vertex form: $y = f(x) = x^2 - 2x - 24 = (x-1)^2 - 25$. What? How did I find the vertex form? First, these two forms can be checked to see that they are the same quadratic function (please check!!). The vertex form is named this way because you can 'see' or identify the location of the vertex: (1, -25). How did I find this vertex? Location of the vertex as an ordered pair is given by the formula: $\left(-\frac{b}{2a}, f(-\frac{b}{2a})\right)$ What? Another way to write it is: $x = h = -\frac{b}{2a}$, $y = k = f(-\frac{b}{2a})$. Since $a = 1$, which is positive (that is, $a > 0$), then the parabola 'opens up' or is a 'touch down' or looks like a 'valley'. For this situation, the vertex, (1, -25) is the **minimum/lowest** point on the parabola when plotted on the Cartesian coordinates. Another way to interpret the vertex is: the minimum value of the function is -25 and it occurs at $x = 1$. So, the value of the function can never be less than -25. If $a < 0$, then the parabola 'opens down' or 'is not a touch down' or 'is a hill', and the vertex is a **maximum/highest** point on the parabola when plotted.

Let's 'plug and chug' by finding the location of the vertex: $x = h = -\frac{-2}{2(1)} = 1$ and $y = k = f(1) = (1)^2 - 2(1) - 24 = -25$. Please be careful here with the arithmetic!! So the vertex is (1, -25). Based on the vertex form shown above, $h = 1$ and $k = -25$. So since $a = 1$, $f(x) = 1(x-1)^2 - 25 = (x-1)^2 - 25$.

6.2 Characteristics or Important Facts of the Quadratic Function

Here is the summary of the results obtained for $y = f(x) = x^2 - 2x - 24$ from the previous Section 6.1:

- The quadratic function 'opens up' since $a > 0$, as $a = 1$;
- The vertex is (1, -25);
- The quadratic function crosses the y-axis at $y = -24$, or the point on the parabola is (0, -24) or called the y-intercept.

Based on these three 'facts', a sketch of the parabola can be made with some degree of accuracy. It does not have to be a perfect replica, but a good representation of an actual one will suffice. The sketch is shown below in Figure 6.1.

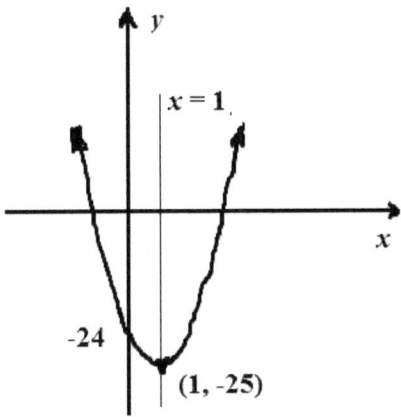

Figure 6.1: Sketch of $y = f(x) = x^2 - 2x - 24 = (x-1)^2 - 25$

Let's draw a vertical line through the vertex. We see that this vertical line through the vertex 'splits' or divides the parabola into two equal 'pieces' (equal left and right sides from this vertical line). See also Figure 6.1. The equation of the vertical line is given by $x = 1$ (do you remember equations of lines, specifically vertical lines?)

Fact #4: The vertical line, $x = h = 1$, is called the ***axis of symmetry*** of the parabola.

Practice Problem Set #12

Sketch the quadratic function given by $y = f(x) = -2x^2 + 2x + 3$ by labelling: the vertex, the x-intercept, the y-intercept, and the axis of symmetry. Convert the standard form to vertex form.

Also, for $y = f(x) = -(x+4)^2 + 3$, sketch $f(x)$ and label the vertex, the x-intercept and the y-intercept, and the axis of symmetry. Convert this function into standard form.

6.3 The Quadratic Equation

Quadratic this, quadratic that ... it can be so confusing!!

There is the quadratic expression, the quadratic function, and now, the quadratic equation? What? I guess I am trying to confuse you? No. Just letting you know that there are differences, and you need to be careful and be aware of the distinctions between the three.

Let me explain what a quadratic equation is. Based on the sketch in Figure 6.1, there are two points on the x-axis where the parabola crosses. Let's find one of these two locations. To find these two locations, the *quadratic equation* is needed to find these two locations. To achieve this, set $y = f(x) = 0$. Why? Any y-value along the x-axis is equal to zero, and we want to find the two locations on the x-axis where the parabola crosses or when $f(x) = 0$. Let's find a value of x that will make $x^2 - 2x - 24 = 0$

So, in general, the quadratic equation is given by $ax^2 + bx + c = 0$. Yes, all the hype and confusion, and that's all this is!

Observation: Note what we just did here. We see that physically, the parabola crosses the x-axis at two locations. To find these two locations, we solve an equation by setting the function equal to zero. That is, we converted a physical problem into a mathematical problem. This is called 'applied mathematics', a branch of mathematics where the mathematics is applied to areas in, for example, biology, economics, earth quake engineering, and digital communication.

Let's continue. We can guess a bunch of x values, say $x = 0, 1, -1, 2$, and -2, but these values do not satisfy the equation. There must be a better method to exactly determine the two x values. After some time, we notice that when $x = 6$, $y = f(6) = 6^2 - 2(6) - 24 = 36 - 12 - 24 = 0$. So, $x = 6$ satisfies the quadratic equation. Check it again!! We were lucky to guess and find a value of x that satisfies the quadratic equation.

Fact #5: As an ordered pair, (6, 0) is a point on the parabola or this point is where the parabola crosses the x-axis. This ordered pair, as you may know, is also called the x-intercept of the quadratic function. Based on the sketch provided above in Figure 6.1, this parabola crosses the x-axis at another location. Let's find the other x-intercept.

Rather than guessing for x-values as before, there must be a better way. Yes, there is, and it is through factoring – the very important 'f-word'! If the quadratic expression can be factored, it can make life of a math student very easy. So, need to solve the quadratic equation given by $x^2 - 2x - 24 = 0$ via factoring. Can the left side be factored? Yes. In fact, $x^2 - 2x - 24 = (x+4)(x-6) = 0$. So when $x = -4$ or $x = 6$, $x^2 - 2x - 24 = 0$. Please check (as always)!! Both $x = 6$ and $x = -4$ are called the _**zeroes**_ of the quadratic function because at those two x-values, the value of the function or the y-value is 0. So, $x = 6$ is one _**zero**_ and $x = -4$ is the other _**zero**_ of the same quadratic function. From these zeroes, we can say (-4, 0) and (6, 0) are the x-intercepts. Great! The sketch provided above can be better represented by labelling these two additional points.

Note: If we directly factored the quadratic expression (the left side of the quadratic equation), we would have by-passed the guessing portion, and therefore we would have saved us time, the precious time that you and I don't have much of.

Figure 6.2 shows the parabola with the two x-intercepts labeled in the figure.

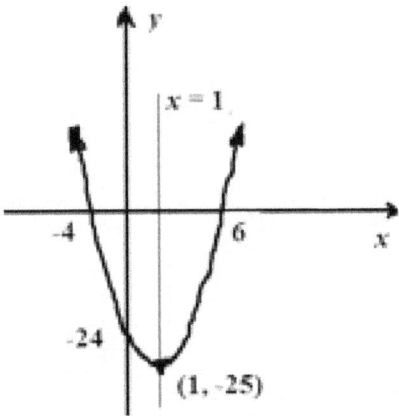

Figure 6.2: Sketch of $y = f(x) = x^2 - 2x - 24 = (x-1)^2 - 25$

Fact #6: As we just discussed, the parabola crosses the x-axis at $x = -4$ and 6. From the axis of symmetry, $x = 1$, we see that on the x-axis, these two zeroes are both 5 units away from the axis of symmetry. This is not surprising since the axis of symmetry splits (or equally divides) the parabola into two equal parts/pieces. In fact, any two x-values on the parabola for a given y-value, are the same distance away from the axis of symmetry.

Fact #7: Alternatively, the quadratic *formula* could have been used to find the solution to the quadratic equation $x^2 - 2x - 24 = 0$. Before, I used to always use the quadratic formula although factoring would have been simpler and better (that is, efficient). Also, from my experience, I made plenty of arithmetic errors when using the quadratic formula. In addition, at times, I used to write the quadratic formula incorrectly which does not help (when you're in a hurry). If the quadratic expression cannot be factored 'easily', the quadratic formula should be used <u>*as a last option*</u>, but please be careful with the arithmetic if and when you are using it. Afterwards, as always, please check your answer(s).

So here is the quadratic formula: $x = \dfrac{-b \pm \sqrt{b^2 - 4ac}}{2a}$ which provides the solution to the quadratic equation. So for our problem $a = 1$, $b = -2$, and $c = -24$, and plugging these values in *correctly*,

$$x = \dfrac{-b \pm \sqrt{b^2 - 4ac}}{2a} = \dfrac{-(-2) \pm \sqrt{(-2)^2 - 4(1)(-24)}}{2(1)} = \dfrac{2 \pm \sqrt{4+96}}{2} = \dfrac{2 \pm \sqrt{100}}{2} = \dfrac{2 \pm 10}{2} = -4$$

or $x = 6$. This is what we found when we solved the quadratic equation by factoring.

Advice: I personally do not like using the quadratic formula, but it was an 'easy' way for me to avoid factoring. With practice, I found out that if a quadratic expression in the equation can be factored, then the quadratic formula should be avoided. Whatever method you choose, do it correctly, and afterwards, always check whether your solution satisfies the quadratic equation.

Practive Problem Set #13:

Please sketch:

a) $f(x) = -3x^2 + 4x - 9$; and
b) $g(x) = \tfrac{2}{3}x^2 - 7x + 1$

by locating the vertex, the x- and y-intercepts, and the axis of symmetry for each quadratic function. Also, rewrite the standard forms into vertex forms.

Chapter 7
Basic Trigonometry: SOH-CAH-TOA

I have heard and still hear this many times from my classmates: "I hate trig!!" I don't know why they hate it. For me, I do not hate it nor do I like it. I took trigonometry because it was a requirement for my Calculus AP course. In trigonometry, you need algebra and geometry to understand trigonometry. You remember what I said at the beginning? Topics in math build from previous math skill-set. Life!

7.1 The Sine, Cosine, and Tangent or SOH – CAH – TOA

The acronyms from the above title are used to determine the three fundamental trigonometric values: the sine, the cosine, and the tangent of an angle.

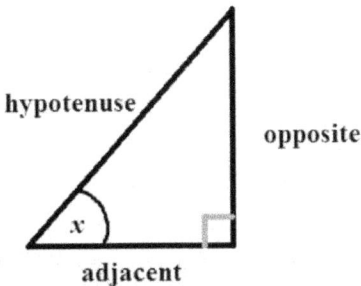

Figure 7.1: Right-Angle Triangle

In mathematics, the 'sine' of an angle x (sin x for short) is a real number/value. It is calculated by using a right-angled triangle shown in Figure 7.1. To calculate sin x, we use this triangle and **define** sin x to be the *length* of the 'opposite' side (from angle x) of this right-angled triangle *divided* by the *length* of the hypotenuse. This is how sin *x* is calculated ... period! A group of mathematicians got together and said this is how we shall calculate the sine of an angle. So we have SOH as the acronym for **sin *x* = opposite/hypotenuse**. Similarly, cosine of the angle *x*, written cos *x*, is calculated by the **definition**, **cos *x* = adjacent/hypotenuse**, and tangent of angle *x*, written tan *x*, is calculated by **tan *x* = opposite/adjacent**. Just remember that these three trigonometric values are **merely definitions** or how you would use the lengths of the right-angled triangle to calculate each of the three basic trigonometric values.

Now, let's do an example to illustrate what those mathematicians were talking about.

Example #12: Find sin *x*, cos *x*, and tan *x*, and sin *y*, cos *y*, and tan *y* based on the right-angled triangle given below:

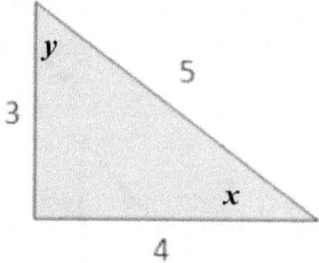

Figure 7.2: Right-Angled Triangle for Example #10

We know from the definitions of the three basic trigonometric values, it is just a matter of using the SOH, CAH, and TOA rules/definitions/formulae. So, it is just plugging in the two lengths of a right-angled triangle to find these trigonometric values. In order to find sin *x*, use the sin *x* = opposite/hypotenuse rule. We know that the length of the opposite side to angle *x* is 3, but we don't know the length of the hypotenuse. So I guess, we cannot determine the value of sin *x* since we don't know it. I am sorry that I wasted your time!! Wait, let's think about this! We *can* determine the value of sin *x* since we can calculate the length of the hypotenuse! Well, what kind of a triangle is given in Figure 7.2? It is a right-angled triangle!! So what do you know about the relationships between the length of the hypotenuse and the other two lengths of this right-angled triangle? Remember the Greek mathematician Pythagorus? Specifically, the Pythagorean Theorem? We use the Pythagorean Theorem to find the length of the hypotenuse!

So let's assume in general, for a right-angled triangle, the length of the hypotenuse is c, and the lengths of the other two sides (also called the *legs* of the right triangle) are given by a and b. The theorem says: $c^2 = a^2 + b^2$. Since the legs of the given triangle are provided, that is, say $a = 3$ and $b = 4$, $c^2 = 9 + 16$, which means that $c^2 = 25$ or $c = \pm\sqrt{25} = \pm 5$. So we choose $c = 5$ since a length of anything must be positive!! Therefore, the length of the opposite side to angle x is 3, the adjacent side 4, and the hypotenuse 5. We are now ready to answer the first part of the question.

So, based on the definitions, $\sin x = \dfrac{3}{5}$, $\cos x = \dfrac{4}{5}$, and $\tan x = \dfrac{3}{4}$. Similarly, $\sin y = \dfrac{4}{5}$, $\cos y = \dfrac{3}{5}$ and $\tan y = \dfrac{4}{3}$. When you look at angle y, the length of the 'opposite' side is now 4 and the 'adjacent' is now 3, just flipped from when we looked at the lengths from the angle x perspective. The length of the hypotenuse remains unchanged since it is still the longest length of the right-angled triangle.

Practice Problem Set #14: Find sin x, cos x, and tan x from the Figure below.

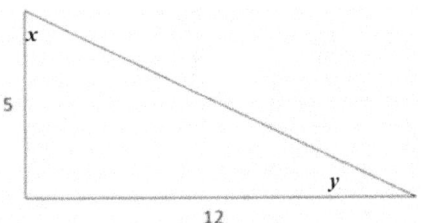

7.2 The Cosecant, the Secant, and the Co-tangent

More trig stuff ... fun!

Mathematician love this stuff and continued with the trig party once they got together. They also defined three additional trigonometric quantities called the co-secant, the secant, and the co-tangent of an angle. These three new quantities are simply 'reciprocals' of the sine, the cosine, and the tangent of an angle, respectively. So, what I mean by a reciprocal is, for example, co-secant of x, written csc x is **defined** by 1/sin x. Yeah, that's it. So, csc x = 1/ sin x. Similarly, secant of x, written, the sec x = 1/cos x, and the co-tangent of x, written cot x = 1/tan x. Fun ... isn't it? Please wake up!!

So based on the example from the previous section, let's find: csc x, sec x, and cot x, and also, csc y, sec y, and cot y. Since each of these values are reciprocals, all you need to do is 'flip' each one of these six quantities. That is,

$$\csc x = \frac{1}{\sin x} = \frac{1}{3/5} = \frac{5}{3}$$

$$\sec x = \frac{1}{\cos x} = \frac{1}{4/5} = \frac{5}{4}, \text{ and}$$

$$\cot x = \frac{1}{\tan x} = \frac{1}{3/4} = \frac{4}{3}$$

Similarly, csc $y = \frac{5}{4}$, sec $y = \frac{5}{3}$, and cot $y = \frac{3}{4}$

7.3 Radian Measure when using Angles

Thus far, I have shown you that when evaluating trigonometric values, the angle used is in degrees. You will discover later when you take math courses in high school that instead of using angle (in degrees), a radian measure for an angle is most common, especially, when you take Calculus. Converting an angle from degrees to radians (and vice-versa) is quite simple, which is like converting inches to feet, or meters to kilometers. The conversion is as follows: 360° = 2π radians or 180° = π radians or 90° = $\frac{\pi}{2}$, etc. Do you see how the other conversions were obtained based from the first one? We can also see that $1° = \frac{\pi}{180}$ radians or $\frac{180}{\pi}° = 1$ radian. So, $30° = \frac{\pi}{6}$ radians, $45° = \frac{\pi}{4}$ radians, and $60° = \frac{\pi}{3}$ radians. We see that in the first column of Table 7.1, the radian measures of each of the angles are also listed.

Practice Problem Set #15

Convert the given degree measures into radian measures:
15°, 50°, 90°, 135°, 225°, and 415°.

Also, please convert radian measure into degrees:
$\frac{\pi}{2}, \frac{3\pi}{4}, \frac{\pi}{12}, \frac{3\pi}{2}, \frac{3\pi}{8}, \frac{3\pi}{2}, \frac{4\pi}{3}, \frac{5\pi}{2}$, and $\frac{6\pi}{5}$.

7.4 The Special Right-Angled Triangles

There are two special right-angled triangles that you need to know, especially if and/or when you take Calculus. They are called the 'special' 30° – 60° – 90° triangle and the 45° – 45° – 90° triangle. For these two right-angled triangles, you will be required to know, say sin 60° or tan 45°, by memory or by constructing one of these triangles quickly.

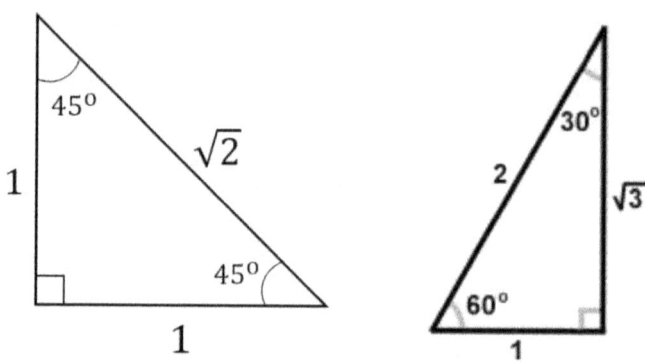

Figure 7.3: The Special 45° – 45°– 90° and 30° – 60° – 90° Triangles

The two special triangles shown in Figure 7.3 have been assigned 'standard' lengths to the sides of these triangles. We will be using these standard lengths throughout this chapter.

Based on these two special right-angled triangles, the trigonometric values are listed in the following Table 7.1:

x	sin x	cos x	tan x	csc x	sec x	cot x
45° or $\frac{\pi}{4}$	$\frac{1}{\sqrt{2}}$	$\frac{1}{\sqrt{2}}$	1	$\sqrt{2}$	$\sqrt{2}$	1
30° or $\frac{\pi}{6}$	$\frac{1}{2}$	$\frac{\sqrt{3}}{2}$	$\frac{1}{\sqrt{3}}$	2	$\frac{2}{\sqrt{3}}$	$\sqrt{3}$
60° or $\frac{\pi}{3}$	$\frac{\sqrt{3}}{2}$	$\frac{1}{2}$	$\sqrt{3}$	$\frac{2}{\sqrt{3}}$	2	$\frac{1}{\sqrt{3}}$

Table 7.1: Trigonometric Values for the Two Special Right-Angled Triangles.

You should try to memorize this table, but if you cannot, just remember these two special triangles, and the basic lengths of the sides of these triangles. Then knowing the three basic trigonometric definitions, these sets of trigonometric values can be easily obtained. I am not good at memorizing, so I would often resort to sketching the right-angled triangles and find the trigonometric values using the definitions.

7.5 The C – A – S – T Rule

Thus far, we have looked at values of sine, cosine, and tangent at angles less than 90° or π/2. Since angles can be greater than 90°, we need to determine how the trigonometric values are calculated when angles are greater than 90°. To achieve this, we are going to need the Cartesian coordinate system as a 'back-drop' to determine how the values of sine, cosine, and the tangent at these angles are calculated.

Example #13: Evaluate sin (135°) or $\sin\left(\frac{3\pi}{4}\right)$. In doing so, we need to establish how 135° will be constructed. Please see Figure 7.4. Also, we need some basics learned from your geometry class.

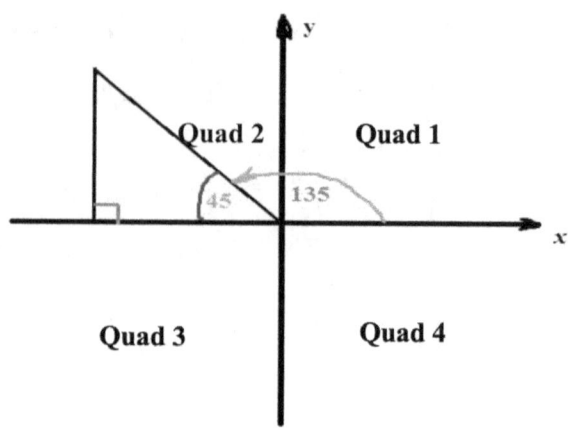

Figure 7.4: How a 135° Angle is Constructed

To illustrate how angles are constructed on the Cartesian coordinate system, place, say, a long edge of a ruler on the positive *x*-axis with the left short edge placed at the origin. The 'convention' or rule used for a *positive* angle is when you pivot the left edge of the ruler about the origin and rotate it counter-clock wise (CCW) 135° away from the positive *x*-axis. As a result, the edge of the ruler now is located within the 2nd Quadrant of the Cartesian coordinates as shown in Fig. 7.4 above. That angle (135°) can also be seen as an angle 45° relative to the *negative x*-axis. This is because, as you may know from geometry, *supplementary* angles add up to 180°. Now, create a right-angled triangle by dropping a perpendicular line (an altitude) from the tip of the left edge of the ruler onto the (negative) *x*-axis. We just created an imaginary 45° – 45° – 90° triangle with the long edge of the ruler being the hypotenuse. We will now use this (imaginary) right-angled triangle to find sin (135°). Since the triangle is located in the 2nd quadrant, we know that the *x*-values are *negative* and the *y*-values are *positive*. So the length of the opposite side from the 45° angle is (positive) 1 (one) or the tip of the opposite side is located on the positive side of the *y*-axis, and the *length* of the adjacent side is 1, *but we say -1 since the tip of the adjacent side is located at* (-1, 0).

So by using the definition of sine of an angle, sin(135°) = opposite/hypotenuse = $1/\sqrt{2}$. For cos(135°) = $-1/\sqrt{2}$, and tan(135°) = $1/(-1)$ = -1. Before when the angles were acute, all three trigonometric values were positive, but in this case, one out of the three trigonometric values are positive (namely, sin(135°) = $1/\sqrt{2}$) with the other two being negative values.

Advice: The strategy here is when asked to find the trigonometric values for angles greater than 90°, create a right-angled triangle with the acute angle (obtained from the idea of supplementary angles), and determine which Quadrant this right-angled triangle is located.

Let's look at a case when a 'negative' angle is used. Yes, negative angles do exist (in mathematics). A 'negative' angle is defined as the angle created with that same ruler that we used before, but now that ruler is rotated clock-wise (CW) away from the positive x-axis (rotated below the x-axis). Mathematicians define negative angles this way. In fact, for example, -30° is the *same* angle as 330°. Yes. That is, if you rotate the ruler 330° (which is CCW), you will end up in the same position as if you were to rotate it 30° CW, which is -30°. Cool isn't it?

So let's find tan (-30°) and the other two trigonometric values that goes with that angle.

Once again, we place one end of a ruler at the origin, and rotate/pivot that end 30° CW. We see that edge of the ruler dips below the positive x-axis and it is now located in the 4th Quadrant. At the end of the ruler, a perpendicular line (altitude) is created to meet at the positive x-axis. We now see a 30° – 60° – 90° triangle within the 4th quadrant. We know that within the 4th quadrant, x-values are positive, and y-values are negative. So as a result, the length of the adjacent side of -30° is √3 and the length of the side of the opposite side is 1, but since the tip of the ruler falls below the x-axis, the length of the opposite side is -1. So, using the definitions, tan (-30°) = opposite/adjacent = -1/1 = -1. Similarly, sin(-30°) = -1/2, and cos(-30°) = ½. We see that in this case, the cos(-30°) is a positive value, and the other two trigonometric values are negative.

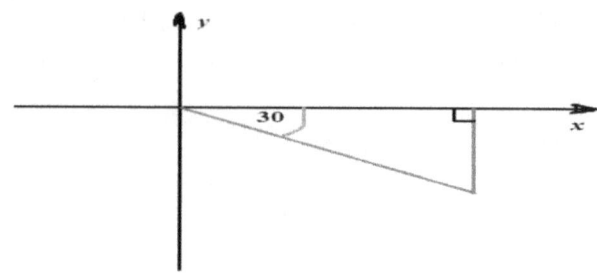

Figure 7.5: Negative-Angled Trigonometric Values

Let's recap what we found based on the two examples.

1. When the angle is in the 1st Quadrant (that is, acute), we found that from the previous section, that **A**ll three trigonometric values are positive;

2. When the angle is in the 2nd Quadrant, **S**ine of the angle is positive and the other two trigonometric values are negative;

3. When the angle is in the 4th Quadrant, **C**osine of the angle is positive and the other two trigonometric values are negative; and

Problems Problem Set #16

4. What happens when the angle is in the 3rd quadrant? What can you tell me about the signs (positive or negative) of the three trigonometric values? Find: sin (240°), cos (240°), and tan (240°)? Hint: The **C – A – S – T** acronym.

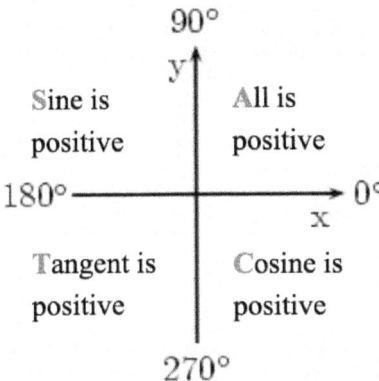

Figure 7.6: Quadrants where the Trigonometric Values are Positive – CAST Rule

Figure 7.6 summarizes the results of the trigonometric value(s) that is/are positive based on where the angle is located in the Cartesian coordinates.

The tangent of an angle within the third quadrant is positive with the other two trigonometric values being negative. You should check!! Thus, the C – A – S – T rule!

Chapter 8

Closing Remarks

You made it.

Congratulations. I hope that you are proud of yourself because most of your friends will not share the same pride in you finishing this workbook. With this book, I wanted to make the subject of math and the introduction to topics that you have done previously or will encounter later in school just a little bit more enjoyable—so I hope that you found it a satisfactory experience. If you didn't, well, it is a math book after all.

I have presented the topics that I find to still be relevant, as well as providing you with my personal perspectives on common mistakes and how to correct them. I am not an expert, so I believe that my experience will be more useful to those still learning. Because I am not an expert, I am a master in the art of mistake-making. You will hear this so much that it might lose meaning, but making mistakes are crucial to progress. So if this book is to teach you anything, it is to embrace your mess-ups. In recognizing these typical errors, we can ensure that they are avoided. Though it is important to work with your peers initially, working independently on these solutions will make the ultimate difference. So please, spend some quality time with yourself, maybe get a snack, put on some music, and keep practicing.

In writing this book, I have come to the realization that although I have had limited exposure to mathematics, the importance of this subject continues to reappear. Going back to what I did two-three years ago, reinforced my understanding. Sure, it may appear that in math you just follow a set procedure or a recipe, but I found the critical thinking component to be fascinating. It forces you perform efficiently and think logically.

The more I learn about mathematics, the more I see how powerful a tool it is in understanding the things around us. For instance, with the cell phones we all carry today. The technology needed to create and operate a cellphone is based upon the basic and complicated fundamentals of mathematics. How can we press a button and turn on a screen of light? How can we use the heat out of fingers to interact with these screens? How can we put our phone on silent and feel it physically buzz? Math. On another topic, how can we most efficiently mow our lawns? Well, my dad and I actually wrote a paper on how to most effectively cut your grass. We analyzed the time it takes to cut your entire lawn based on how one typically cuts their lawn (there is a systematic pattern used to cut the lawn). In the analysis, I was surprised to use the very skills I had learned from middle and high school algebra, which played an important role in determining and assessing the best cutting method/pattern. Who asked for this analysis? No one. Why did I take part? Why not? But I'm sure it may be of use to someone one day. If not, then at least my lawn will always be well manicured and look nice.

Enough of me.

On a lighter note, I have provided below twelve renowned and celebrated mathematicians and scientists. I want to share with you the people I believe made some of the most significant contributions not just in the field of mathematics and sciences, but in the world as a whole. Pictures of them are attached (which are real) and their famous reviews/quotes (for humor) about the work I have done in this book are provided. Yup, they really reviewed my book—totally not written by me. 😊

Practice Problem Set #17:

Please find more about these prominent people, namely, when and where they were born, what contributions they made in their area, and any non-mathematical and/or non-scientific areas that they are also known for. There are clues (in each quote) where each person has made significant contributions in their area. I am sure you have heard some of them, and if not, find out who they are!

Have fun doing mathematics and please continue to do so!

Important People
That endorsed me

"There is a 'ring' to this book that I find fascinating! I will apply the work presented in this book to my 'field' of work. Great job Josh!!"

Emmy Noether

"Good. The 'special' attention to detail that Joshua focuses in is significant, but only 'relative'ly gratifying."

Albert Einstein

Rene Descartes

"Someone get me the 'coordinates' to Josh's mind. Brilliant. Revolutionary. A must-read book for young students today."

Sofia Vasilyevna Kovalevskaya

"I cannot 'partially differentiate' genius from the work of Mr. Lee. This book offers something great, standing out from the rest of the math books out there for middle and high school students. The 'mechanics' of algebra are concise."

"I am 'set' to buy this Math book. The quality of this book is im'measurable'! Oh my god! Joshua B. Lee is the guy!"

Georg Cantor

"Joshua B. Lee is passionate and I would have been better off if this book was around when I was in middle school!! Too bad I'm dead. This math book will 'transform' your perspective of math and of life!"

Pierre-Simon LaPlace

Majorie Lee Browne

"Josh, your book appeals to me because it promotes conscientious efforts within the 'classical group' of middle and high school students. Wish a book like this was available back in Tennessee when I was your age."

Simeon Poisson

"I will 'probably' go to see Mr. Joshua B. Lee and ask him questions about this well written math book. If not, maybe we can get some burritos and talk about something even more fun like probability."

Bernhard Riemann

"You must read this book. After reading this, I have developed a 'hypothesis' I am still testing – Joshua B. Lee is much cooler and better-looking in person."

Brooke Taylor

"Profound! I'm going to go buy this math book after I get changed. I'm thinking about some skinny jeans and crocs. I'm not joking. I'm very 'ser-i-es' (serious)!"

Katherine Johnson

"Josh, your 'trajectory' towards greatness is obvious. I can see that your peers will 'orbit' around you! It seems like you are right there with me when you explain things in the book. Great job sir!'

Elbert F. Cox

"You are the 'first' to personalize your understanding of algebra to middle school students in this well written book. This, in fact, will make a big 'difference' to them, and instill confidence that they need and deserve. I see a 'Big Red' sign that flashes 'Please read this very informative and comical book!'"

Math Guide

www.ingramcontent.com/pod-product-compliance
Lightning Source LLC
Chambersburg PA
CBHW070809220526
45466CB00002B/615